RENEWALS 691-4574

DATE DUE

DEC 09			
DEC 08			
MAR 1 4			

Demco, Inc. 38-293

An **ama** Management Briefing

Quality Assurance: A Program for the Whole Organization

Victor J. Goetz

A Division of American Management Associations

Library of Congress Cataloging in Publication Data

Goetz, Victor J
    Quality assurance : a program for the whole
organization.

    1. Quality assurance. I. Title.
TS156.6.G63   658.5'6   78-5336
ISBN 0-8144-2221-7

®1978 AMACOM

A division of American Management Associations, New York.

This Management Briefing has been distributed to all members enrolled in the Manufacturing Division of the American Management Associations. A limited supply of extra copies is available at $5.00 a copy for AMA members, $7.50 for nonmembers.

First Printing

Contents

About the Author

Victor J. Goetz is manager of quality planning and development for the Warner Lambert Company, where he is responsible for providing worldwide consulting, auditing, and training services related to quality policy, management, and practices. Prior to his present position he was director of administrative services for STAT-A-MATRIX, Inc., an international consulting and training organization specializing in the development of quality assurance programs.

Mr. Goetz has over 25 years of diversified industrial and government experience, including various management and engineering activities concerned with manufacturing, purchasing, quality assurance, quality control, and inspection. He has developed and conducted engineering and quality assurance courses at several community colleges in Washington State, for the American Society for Quality Control, and for American Management Associations.

He is a registered professional engineer, a senior member of the American Society for Quality Control, an ASQC certified quality engineer, past chairman of its Seattle section, and current vice chairman of the Metropolitan section. He holds a Bachelor of Science degree from Cornell University and has done graduate work in management at the University of New Mexico and the University of Washington.

Mr. Goetz is the author of *The ASME Quality Assurance Manual* and is a contributor to three other books on quality assurance. Several of his papers presented before the ASQC have been published by that organization.

Introduction

RAPIDLY advancing technology, increasing complexity of operations, and growing competition in the marketplace have made modern industry painfully aware of the necessity to provide, as economically as possible, products and services that satisfy customer requirements. Fortunately, quality assurance concepts and techniques have been evolving in a way that facilitates achieving these quality and cost objectives.

Formerly, the maintenance of quality was thought to be something that was done after, and apart from, the more important work of engineering and manufacturing. It was considered an activity that would be unnecessary if engineering, and particularly manufacturing, always did their job, and there was often a lingering hope that when current problems were solved the function could be discontinued. The cost of the quality function was often thought of as a direct reduction in profits. To shop supervisors, charged with maintaining a production schedule, and to workers paid on an incentive basis, the function was an obstacle to be circumvented whenever possible.

For their part, people involved in the quality function felt that they were burdened with problems created by others, that their contributions were unnoticed and unappreciated. They saw themselves as caught between the push of manufacturing to get the product out the door and the insistence of marketing that no faulty material reach the marketplace. Their position was all the more difficult, since in many cases the function reported to either manufacturing or marketing.

Most of these difficulties can be traced to an inadequate conception of the quality function—it was viewed as a kind of policing activity. The assurance of quality is in fact an important aspect of all activities that enter into the matching of customers' needs or desires with company-produced products and services. Quality assurance begins with the customer's need. That need must then be effectively translated into design documents and manufacturing processes. Unless it is properly designed and manufactured, a product cannot and will not meet the customer's requirements. Consequently, companies must develop and implement the systems by which quality is assured, beginning with the customer's order and continuing through all the intervening steps until the product is delivered and put into use.

Quality assurance is important to all kinds of organizations. To some degree it touches almost every employee, and it extends to suppliers and subcontractors. Yet no company is precisely like any other. Each must develop and implement its own plan of quality assurance. How, then, is this to be done? The key lies in determining the principal elements of a system that will satisfy customer requirements as economically as possible.

This briefing emphasizes a comprehensive approach to quality assurance. It will explore basic concepts and indicate how a cost-effective quality assurance program can be designed, documented, implemented, and controlled.

BASIC TERMINOLOGY

The meaning that an individual employee attaches to quality assurance depends on the context in which that employee experiences it. He or she may associate it with conformance, regulations, or specifications, but from the overall corporate point of view, it is basic to the idea that we are in business to satisfy our customers, the users of our products and services. Quality assurance is an important and indispensable part of the process. To put it more formally, *quality assurance* is made up of all the planned and systematic actions that provide confidence to an organization that its product or service will give satisfaction when it is put into use.

Quality control, strictly speaking, is the process by which we maintain the characteristics of a product or service within preestablished ranges of value. We do this by measuring the characteristic as it is found in the product or service and comparing it with a known standard. That is to say, quality control refers to those portions of the quality assurance program that are concerned with the physical characteristics of a product or a service and that measure and maintain (or regulate) it so that it conforms to predetermined requirements.

Quality assurance does not begin with quality control. It goes back to the design of the product, where customer performance requirements must be realistically defined, taking into account both customer expectations and shop capabilities. If the product is overengineered so that stated tolerances cannot routinely be obtained in the shop, there will be great difficulty in attempting to control quality within the specified limits. Quality assurance really begins in marketing, where customer requirements and specifications should be defined and be related to manufacturing capability.

Other terms used in connection with quality assurance are *inspection, test,* and *auditing.* As used here, inspection refers to the act of examining the product for the purpose of accepting or rejecting it. Inspection may be visual or instrumental. In either case it measures specific, isolated characteristics. Testing, where the term is applicable, refers to an evaluation of the performance of a product. Auditing, on the other hand, is designed to evaluate the system that produced the product.

QUALITY ASSURANCE FUNCTIONS

Quality assurance, in its broadest sense, emphasizes doing the job right, whether the task is in marketing, engineering, or manufacturing. In a more specialized sense, the term refers to the programs, the technical discipline, and the organizational unity established to verify and document the satisfactory completion of work. In this context, the term "quality assurance" (as a technical specialty or as a formal organization) describes a staff support function designed to assist management in realizing its overall goal of high-quality performance of equipment, procedures, and personnel.

Historically, quality assurance as an accepted discipline has been associated with manufacturing and construction, where it operated as a separate inspection function. It is usually identified with systems of checks, audits, inspections, and other forms of verification that can be carried out before the product is put into service. The nature of manufacturing and construction is such that time is usually available for inspecting the product and correcting any defects that may exist.

In contrast, deficiencies in process industry operations have an immediate effect, and it is important that the operations be monitored on an essentially continuous basis. Instrumentation for monitoring, controlling, and actualizing safety systems, as well as observations and responses by the operating staff, is extensively used for this purpose. Continuous monitoring of the operation is the most significant part of quality assurance in these cases.

QUALITY ASSURANCE PROGRAM

The marketing of any product requires the performance of certain predetermined activities to design, manufacture, and sell the product, and to maintain or improve the physical plant. A description of the means by which these activities are to be accomplished can often be found in policy statements, procedures, drawings, specifications, and instructions, which, taken together, essentially describe the quality assurance program itself. The development of these documents is a dynamic and iterative process involving preparation, approval, compliance, debugging, and prompt resolution of any inadequacies discovered in the course of the operation. If not formally provided, a form of quality assurance gradually evolves. Whether they are specifically documented or not, quality assurance activities are based on what the particular enterprise perceives as necessary to produce a satisfactory product, while complying with government regulations and industry standards. It follows that most organizations maintain some kind of control over the quality of their operations and products, although they do so with varying degrees of formality. Some businessmen and entrepreneurs will have difficulty accepting this premise, but a brief review of almost any operation will reveal the presence of quality assurance activities, whether or not they are consciously implemented, documented, or effective.

It is also true that most organizations, regardless of the type of industry, follow basically the same pattern in determining customer needs and providing for their satisfaction. Details may vary, but certain functions must be performed in all organizations. First, a marketing group identifies needs and defines markets. If the organization does not already have the requisite productive capability, then that must be provided. If the market is geared to individually designed products, the corresponding design capability must also be established. Once the market has been identified and the production capability developed, the product must be sold.

Eventually, an outside purchase order or an internal work order starts a complex series of activities including design work (where necessary), material purchases, and the scheduling and completion of manufacturing work. After production and a final verification that customer requirements have been met, the product is shipped to the customer. However, the company involvement does not usually stop at that point, but continues under warranty provisions or in the maintenance of repair and field-service facilities. Each of these activities has a potential for error or failure. It is the task of quality assurance to prevent this potential from becoming realized, and, if it does, to identify the problem at a time when repair or replacement is easiest and cheapest. Finally, quality assurance must provide the feedback that will lead to the elimination, or at least the reduction, of similar losses in the future.

DOCUMENTATION

The quality assurance program comprises substantially all corporate activities whose objective is to maintain established standards. A company that prospers in a competitive market is obviously satisfying its customers' requirements and, whether it knows it or not, has a quality assurance program. However, such a program that has evolved on its own is apt to contain some ineffective or redundant elements, which can be modified or eliminated by analysis and design. Usually, a rationalized and visible system is less expensive and more effective than an informal one that has been in operation for some time.

Even formalized and visible systems can usually be improved and simplified when the operations are carefully studied. Experience indicates that up to 30 percent of the paperwork can be eliminated, since much of the documentation does not currently contribute to improving quality or reliability. Changing conditions and requirements make procedures obsolete on a piecemeal basis, but very often the paperwork is not revised to reflect these changes. In one instance, a major company mounted an in-depth review of more than 6,000 instructions and procedures that had evolved over a period of years. The objective of the review was to simplify the method of doing business, reduce paperwork, and thereby climb out of a mire of oversystemization. The first attempt resulted in a 20 percent reduction, including elimination of a procedure that prescribed use of a company airplane that no longer existed. (Sadly, it had collided with a truck some years before.)

The point is that good procedures can become ridiculous when circumstances change. This is not to condemn procedures, but to emphasize the need for keeping them up to date. We create paper to do a job, legitimately, and then forget about it, sinfully. A well-planned and executed quality assurance program does not have to result in unnecessary paperwork; rather, it should provide methods that help employees get the job done.

The need for documentation is threefold: first, the necessary activities must be planned, and the plan must be communicated to those who will perform the tasks; second, there must be evidence that the required work has been done satisfactorily; third, there must be an auditable trail, so that the cause of failure can be determined and corrective action provided.

ORGANIZATION

Effective quality assurance starts with company policy. A statement from top management, emphasizing that the company is committed to satisfying government, industry, and customer requirements and to developing a quality assurance program, should be included in a quality assurance manual. It is important that top management spell out the authority, responsibility, and accountability of the persons and organizations involved. The quality-related functions of all departments should be identified, but no attempt should be made at this level to specify the form or size of the quality assurance organization. Rather, only a framework that has the complete support of top management should be established.

1

Developing the Quality Assurance Program

MANY organizations feel that they are already operating quite effectively and that a formal quality assurance program would do nothing more than create a great deal of unnecessary paperwork. But when these same companies begin to develop a formal quality assurance program, many of them find that their operations are not as effective as they had thought. They find that the discipline imposed by a quality assurance program improves their operations. On the other hand, some companies overrespond and develop programs that are more elaborate and cumbersome than necessary, thereby confirming their earlier belief that quality assurance is a "paper-creating" system.

In a sense, both perceptions are correct; the best system is the simplest one that will meet the needs of the particular company. Two very important steps in the development of a quality assurance program are, first, the determination of essential needs, and second, a careful examination of what is already being done to satisfy these needs.

ORGANIZING FOR QUALITY

At the outset, a clearly identified person or group should be assigned the responsibility for designing, installing, and enforcing the quality assurance program. This arrangement will provide continuous direction to the program as it moves through these phases. Equally important, it will provide facilities for coordinating the efforts of individuals who are not part of the central quality assurance group, but who will nevertheless play an important part in the development and operation of the system.

For convenience, the process of designing and installing the program can be broken down into the following steps:

Defining objectives
Collecting data
Preparing process flowcharts
Correlating needs and practices
Establishing priorities and scheduling improvements
Installing the program
Monitoring and adjusting the program.

DEFINING OBJECTIVES

It is desirable to define both general and specific objectives to guide the development and operation of the quality assurance program. General objectives that are particularly useful include: (1) assigning responsibility and delegating authority for quality-related activities; (2) designing, documenting, and implementing a program specifically adapted to company needs; and (3) assuring that all concerned personnel understand the program and can effectively perform their related duties. Specific objectives should be developed within this framework, defining what the program should accomplish when it is fully operational. Often it will be useful to identify initial and intermediate goals, but care must be taken to ensure that temporary objectives, although appropriate at the time they were developed, do not become a permanent part of the system and thus limit its effectiveness.

Management will usually have many specific objectives in mind at the outset of the study. Shortcomings in accomplishment, as well as the more obvious costs associated with poor quality, usually account for some of the initial managerial interest in a quality assurance program. They may also be planning to enter some new markets, such as nuclear, military, medical devices or pressure vessels, and need to be assured that their quality program will effectively satisfy special requirements. Other objectives, equally useful, may not be perceived until the quality assurance study is under way, or until the program itself is in operation. During the planning and implementation of the program, consideration should be given to periodic revision of the program's objectives.

COLLECTING DATA

All organizations have existing procedures that govern their operations and affect the quality of their products. Some have very rigid procedures, maintained formally in bound volumes; others operate informally, with instructions communicated by word of mouth. The usual first step in developing a new or improved quality assurance program is to collect a set of the existing procedures. For this purpose, procedures are considered to include policies and instructions or any document that specifies a quality-related act.

When the collected documents are examined, it is not unusual to find that many, or most, procedures are either undocumented or obsolete. It is then necessary to find out what is being done and to prepare a written description of it. Although it will probably be necessary to gather additional data from time to time, data gathering as a preparatory step can generally be considered complete when the documents provide a coherent story of current quality-related procedures. At this point, the descriptions need not contain all the details, but they should clearly identify all the quality-related activities.

PREPARING PROCESS FLOWCHARTS

It will, of course, be necessary to examine the details of the activity identified during the data gathering phase and to question the necessity or effectiveness of what is being done. One of the best ways of accomplishing this task is by constructing process flowcharts for each of the activities. Usually, the stream of activity, or work flow, is charted by breaking down the action into small steps that can be classified as an operation, a transportation, an inspection, or a storage (or delay). The flow is represented by a continuous line, branching out whenever alternative actions are possible. Symbols are added to the line to show whether at any particular point a specific operation, transportation, inspection, or delay is taking place. A brief descriptive note of each activity is written in next to the symbol. For example, it may describe an operation and the reason for it, show the duration of a delay, or the distance of a transport. Sufficient detail should be included so that the necessity and effectiveness of each step can be evaluated.

Two kinds of flow need to be represented and examined: the first is the flow of material and the sequence of work being performed upon it (including quality checks); the second is the flow of documents and the entry of information, which together control material movement and work activity. Quality assurance is concerned with both these flows. For example, an inspection operation involves an examination of the work piece and a judgment that it is, or is not, up to standard, whereas the paperwork serves as a record of the inspection facts. It may also serve as a release of the work, and as an order for movement to the next operation.

Because quality assurance must be concerned with the flow of both materials and information, it is preferable to show both on the same chart. Although the preparation of charts and their analysis may sometimes seem tedious, there are substantial rewards in eliminating unnecessary and redundant work and introducing quality checks at a time when any necessary corrections can be made easily and relatively inexpensively.

When all parts of the quality system have been plotted and the charts arranged in the order in which the activities occur, a careful examination and comparison of charts will reveal gaps and overlaps, which can then be eliminated. Occasionally, outright cross-purpose activities will be exposed, and one or both activities can be eliminated or rationalized. Figure 1 is a simplified illustration of the flow of work and documents associated with the receipt and inspection of incoming material.

CORRELATING NEEDS AND PRACTICES

Each quality operation should be examined to see whether it contributes to satisfying a customer requirement or government regulation. If it does not, it probably should be eliminated. If it does satisfy a requirement, it should be analyzed to see whether it does so in the best and most economical way possible. In some cases, it may be appropriate to discuss possible changes with the con-

Figure I. Flowchart: Receiving inspection.

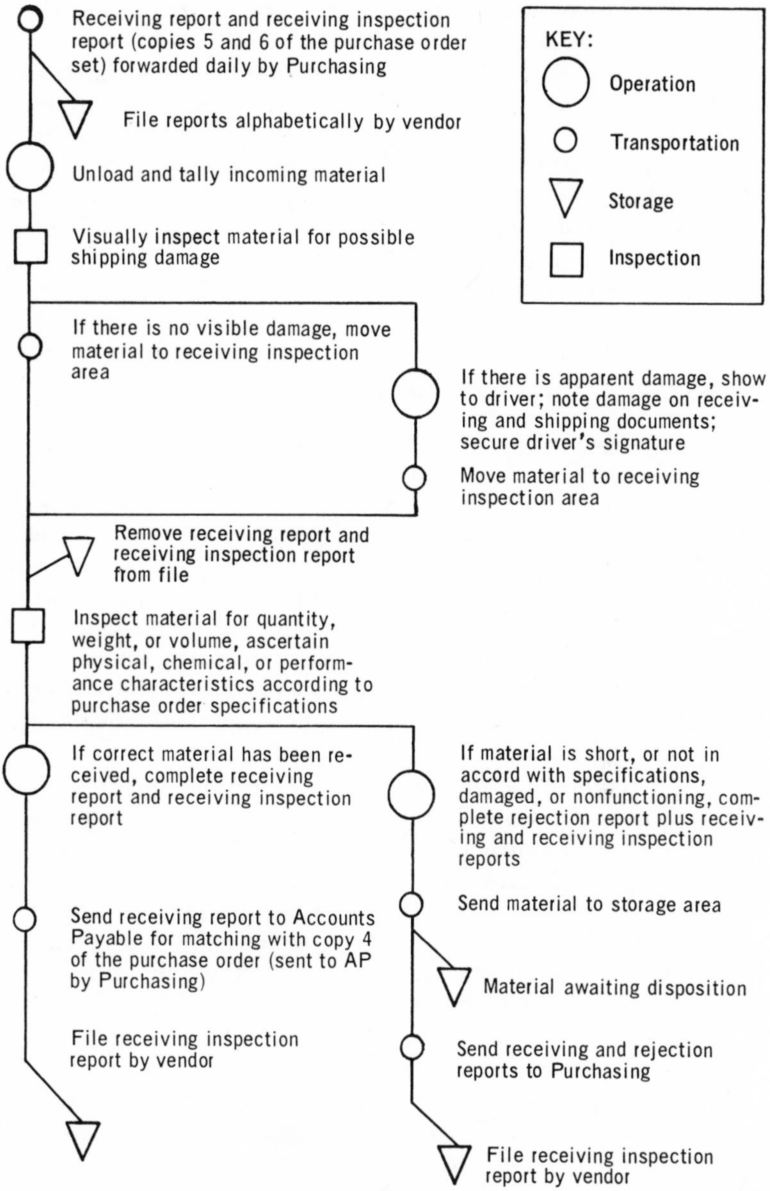

○ Receiving report and receiving inspection report (copies 5 and 6 of the purchase order set) forwarded daily by Purchasing

▽ File reports alphabetically by vendor

◯ Unload and tally incoming material

☐ Visually inspect material for possible shipping damage

○ If there is no visible damage, move material to receiving inspection area

◯ If there is apparent damage, show to driver; note damage on receiving and shipping documents; secure driver's signature

○ Move material to receiving inspection area

▽ Remove receiving report and receiving inspection report from file

☐ Inspect material for quantity, weight, or volume, ascertain physical, chemical, or performance characteristics according to purchase order specifications

◯ If correct material has been received, complete receiving report and receiving inspection report

◯ If material is short, or not in accord with specifications, damaged, or nonfunctioning, complete rejection report plus receiving and receiving inspection reports

○ Send material to storage area

○ Send receiving report to Accounts Payable for matching with copy 4 of the purchase order (sent to AP by Purchasing)

▽ Material awaiting disposition

File receiving inspection report by vendor

○ Send receiving and rejection reports to Purchasing

▽

▽ File receiving inspection report by vendor

KEY:

◯ Operation

○ Transportation

▽ Storage

☐ Inspection

15

cerned customer (for example, when the requirement appears to be unnecessary, overstated, or uneconomical). If the requirement is in fact excessive or unrealistic, changing the requirement can even lead to reduced prices or increased profits.

When a quality assurance system has been fully rationalized — that is, when everything being done is known to be essential and adequate for the intended purpose — management will usually be able to identify the points of corporate strength and weakness. By building on existing strengths and bolstering weak areas, management can tailor the program to the precise needs of the company. Such an approach is cost-effective, because it takes full advantage of existing capabilities.

ESTABLISHING PRIORITIES AND SCHEDULING IMPROVEMENTS

After the needs and opportunities have been identified, a plan of action must be developed to assure that all the necessary work will be done at the right time and in the right sequence. Diagraming essential activities can help accomplish this. Two excellent techniques are milestone charts and network analyses, such as the critical path method. Figure 2 shows a generalized milestone chart, the steps of which can be applied to many different kinds of programs, including quality assurance. Figure 3 shows a simplified network diagram that represents the events and activities involved in developing a product and marketing program.

In comparing the two charts, it should be noted that the milestone chart shows the estimated time for program elements from starting date to date of completion. By looking at the chart, it is possible to tell which activities will be going on at the same time and, when the program is in operation, to see which activities are ahead of schedule and which are lagging behind. On the milestone chart, the elements are represented as though they were independent, but the critical path method shows the interdependence of elements and accomplishment times. The critical activities are those taking the longest time. For example in Figure 3, producing the product is the critical path, and management attention is thus directed toward improving the planned performance time to provide a better match with the shorter time needed to complete the marketing program.

Figure 2. Milestone chart — Quality program plan.

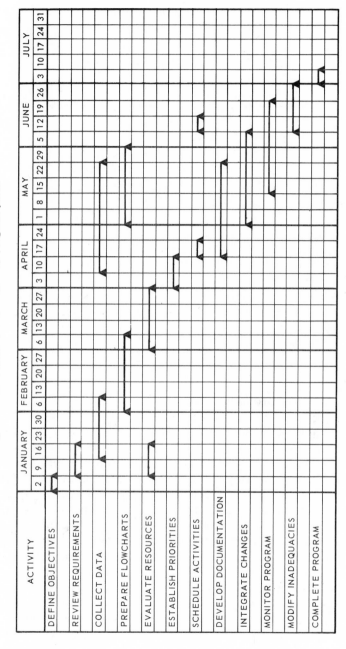

Figure 3. Simplified network diagram showing development of a product and marketing program.

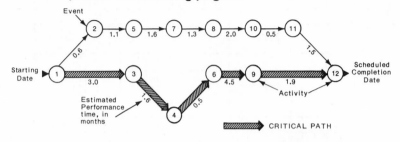

EVENTS	MONTHS REQUIRED	
	To Produce Sales Material	To Produce Product
1. Project authorized		
2. Preliminary marketing evaluation completed	0.6	
3. Product design completed		3.0
4. Design not approved; modifications introduced		1.6
5. Marketing program completed	1.1	
6. Product design approved		0.5
7. Marketing program approved	1.6	
8. Sales and advertising program developed	1.3	
9. Manufacturing started		4.5
10. Advertising copy and layout approved	2.0	
11. Proofs received and approved	0.5	
12. Product and sales material in inventory	1.5	1.9
	8.6	11.5 ——— Critical path

INSTALLING THE PROGRAM

When the evaluation and planning steps have been completed, it should be possible to specify an effective quality assurance program that will mesh closely with the management procedures of the particular company. If the quality program in its simplest form still seems complicated, it will be desirable to design a flowchart of the new procedures as a way of making sure that all foreseeable conditions and circumstances have been taken into account without duplication of effort or backtracking.

Whether or not a flowchart is used, the final and authoritative statement of the program should be made in an operating manual that is understandable to everyone involved in the program. The manual should describe the allocations of authority and responsibility and delineate the procedures to be followed. A manual differs from a collection of informal memoranda in that it has been systematically developed, has nearly universal application, and (until it is revised) specifies precisely what is to be done and by what means.

Usually a manual will be divided into convenient sections, each one covering a specific topic. The division of the quality assurance program into individual topics, and the organization of the topics into a manual, can be done in a number of ways — for example, by using the objectives of the program as the principle of division, with a separate instruction devoted to each objective. Another possibility is to use a list of customer requirements as the organizing principle. Still others include a breakdown by product, customer, or inspection method. A combination of methods may be used as well. In any case, the manual is subdivided to make it easier to use. Therefore, individual instructions should be self-sufficient, so that they may easily be referred to and revised as necessary.

Individual instructions should have a descriptive title, an issue date, and a reference to the individual (or the position) who approved the instruction and is authorized to revise it. It is desirable to give the title page of each instruction a uniform and distinctive appearance and to provide a binder. In the body of the instruction, the "who," "what," "when," "where," and "why," of the required action should be described in clear and simple terms. If there are many instructions in the manual, the first one should describe the quality assurance program in general terms, and also describe the organization of the manual and the provisions for its upkeep.

Before putting the instructions into practice, it is desirable to have the manual reviewed by those who will use it to ensure that it is complete and easily understood. If new or unfamiliar procedures are required, some training of operating personnel may be necessary. If the instructions call for the use of particular forms, these must, of course, be designed and put into stock prior to the beginning of the program. Finally, the program must be initiated on a particular day. To facilitate the transition from old to new methods, the program should be accompanied by adequate advance publicity. Any resistance to change can be overcome if the importance of the program is brought home to each participating employee through such methods as brief employee meetings under the direction of a senior executive. Where this is not practical, a written communication, preferably from the president or general manager, should be used.

MONITORING AND ADJUSTING THE PROGRAM

There are two aspects to what should be an on-going review of the quality assurance program. One is a review of performance. (Is the program being followed?) The other is the review of the program itself. (Is it effective? Can it be improved? Are there easier ways of getting equally good results? Have new conditions developed that are not adequately covered? Has some portion of the program become obsolete?)

Obsolescence is a hazard to all programs. As circumstances change the program is apt to be modified informally, but as the number of changes mount, the formal procedure becomes less and less realistic. Meanwhile, the accumulating informal changes are increasingly cumbersome; they are difficult to remember, different people apply them in different ways, and they are not available for reference or for training new personnel. In fact, a half-formal, half-informal procedure is probably less useful than one that is either completely formal or completely informal.

Thus, to maintain the validity of a program, it is necessary to make changes in a manner that will keep the procedures realistic and available to all interested persons. The exceptions and departures that develop spontaneously should not be repressed by disciplinary means, but rather they should be seen as opportunities for revising the standard program. Employees should be encouraged to report instances in which the program is apparently ineffective. If their suggestions for improvements are genuinely encouraged, they can make a substantial contribution to the successful operation of the program. Nevertheless, all the possibilities for improvement cannot be identified this way, and periodic audits should be made to determine the overall adequacy and effectiveness of the program.

2

Documenting the Program

TO be fully effective, a quality assurance program must be put in writing. In finished form the quality manual provides a comprehensive statement of the philosophy, policies, objectives, and procedures of the program. A statement of philosophy from the chief executive is a good place to begin. Quality program objectives and quality-related manufacturing and marketing objectives should be formulated. These objectives help define the boundaries of the program and the manual's content and structure.

In addition to establishing a firm basis for administering and controlling the quality program, the manual can be an effective sales tool, emphasizing the dedication of the entire company to product quality and showing specifically how desirable quality levels can be attained. Increased customer confidence should result in increased sales and a rising trend in earnings.

Aspects of the quality program that should be considered and formulated include broad statements of philosophy and policy, statements of authority and responsibility, results to be accomplished, coordination of activity, and detailed statements that describe how a particular task should be done. Depending on the size and complexity of the company and of the quality assurance program, it may be desirable to provide for three levels of documentation: policies, procedures, and instructions. When the situation is relatively simple, the levels need not be formally separated, but it will still be desirable to separate the policy, or general guidelines, from the procedural portion of the documentation. To omit this separation often leads to unnecessary or counter-productive action.

POLICIES

Policies describe in general terms what the company wishes to accomplish. In this respect, they resemble objectives. Policies also cover the general rules for decision making when specifying the procedure to be followed is neither possible nor practical. They also help ensure that similar judgments will be made in similar circumstances throughout the company.

PROCEDURES

Procedures generally provide the means for initiating, sustaining, and coordinating all the activities necessary for accomplishing a specific purpose. In the aggregate, they describe the systems in operation in the organization. They document specific responsibilities and authorities, provide for flow of work and information, and regulate the creation and storage of documents and records. They often cross departmental lines, since the business of the company is usually conducted through a network of systems and procedures. For example, in the quality assurance system, procedures might be developed to control design standards, specification and inspection of purchased material, machining and finish tolerances, inspection and test equipment, and scrap and salvage.

Because of the number and variety of actions that need to be coordinated, much of the detail must be omitted and provided for in a separate instruction. For example, a procedure dealing with the inspection of incoming material might call for the inspection of bar stock in line with the provisions of one named instruction, and the inspection of plate stock according to the provisions of a different instruction. Relieved of these details, the procedure can concentrate on coordinating, timing, and sequencing functions and operations.

INSTRUCTIONS

Instructions provide the detailed directions for doing a particular task, and the prescribed method that must be followed. Instructions are essentially control documents, and typically operate within a single administrative area. They are frequently within the jurisdiction of a single supervisor or foreman.

Policies are, in effect, the statement of the long-term, basic orientation of the company, and they seldom require change. Procedures, or the general methods of doing business, do change, but still have a considerable amount of stability. What changes most often is the specific manner of doing a particular task, the initiation of new tasks, and the dropping of old ones. Instructions are relatively easy to change: the areas affected are limited or concentrated, and the facts concerning them can usually be identified and expressed in quantitative terms. The division of the manual into policies, procedures, and instructions thus facilitates review and revision.

When changes have first been made at the level of instructions, the broader effects, if any, can be more easily seen, and the process of updating the related procedures will be simplified. Finally, if a procedure is in conflict with stated policy, the need for changing the procedure or revising the policy will be highlighted. After the policy or procedure has been adjusted, the organization will again have a set of documents that provides overall, top-to-bottom guidance for corporate activity.

A manual can also be used as a device for implementing planned change in management practices. Changes made in policies can be cascaded downward to motivate appropriate changes in procedures and instructions, thus causing unfavorable or unexpected consequences to surface more quickly and at a time when they can be dealt with more easily. Expanding the details of a proposed change by modifying policies, procedures, and instructions, in that order, provides a means of incorporating the operating expertise that is present at various levels in the organization. It also tends to generate a broad acceptance of responsibility for making the changes work.

LANGUAGE, STYLE, AND POINT OF VIEW

Before an idea can be communicated to others, it must be translated into words or diagrams or both, but this seemingly simple process has many pitfalls. For one thing, words have different meanings to different people, largely because of differing associations. A great deal of semantic confusion can be avoided if the author is careful to use words in the way the reader would use them. The author will continually need to ask what the reader has to know at any specific point in order to fully understand what is to follow.

Generally, manuals should be written in the present tense. Documents are intended to be applicable to current circumstances. If the documents are intended to provide a series of steps, some to be taken immediately and some in the future, the present tense should be used.

Ideas should always be expressed in the simplest, clearest language that can be used without creating misunderstandings. Once written, the material should be ruthlessly edited to remove unnecessary or redundant matter and to check for over-or under-generalization. If something has been treated as if it were always true, when in fact it is true only under certain circumstances, that needs to be specified.

In structuring a manual, it is usually better to start with a statement of policy and follow up with the implementing steps. In describing the steps, it is usually easier to follow a scenario, which essentially organizes the presentation on the basis of the action taken by various individuals and, as nearly as possible, in the sequence in which the action takes place. Other information necessary to understand or validate the action sequences (such as the authority of the individual taking the action, or the conditions or circumstances giving rise to the action, or the basis for making a particular decision) can be introduced at the time the action is taken.

FORMAT

The manual serves a variety of needs and purposes whose satisfaction will have some bearing on the format of the documents. The manual has an educational function, providing information and instruction for individuals performing quality-assurance tasks. It serves as a reference book for those seeking authoritative answers to particular questions, and as orientation material for individuals outside the quality assurance group who seek to work more effectively with it. Finally, it provides a source for describing the quality program.

In addition to these utility functions, the manual may serve a persuasive function, particularly when a formal system is first being installed. It can motivate acceptance, cooperation, and good performance, and it can enhance the corporate image and promote sales. All these matters need to be given appropriate consideration when the manual is being designed.

For purposes of image-building and motivation, the binding, printing, separators, and other physical characteristics of the manual should reflect quality. To facilitate its use as a reference volume, the manual should have a table of contents which should be periodically updated.

If the manual is large, section dividers will be useful. The binder should be constructed to facilitate the insertion and removal of sheets. Even so, removing obsolete documents and inserting new ones is often neglected or deferred. Therefore, it will be useful to issue a new table of contents at fairly frequent intervals, indicating the date of the most recent revision next to each document title. Most people will retain and use these sheets even when they do not systematically update the binder.

It is a good idea to put a description of the program, the quality assurance objectives, and lists of specific authorities and responsibilities into a separate document to be filed ahead of the procedural ones. This document will normally be issued by the chief executive. In many cases, it will also be desirable to emphasize top-level support by prefacing the manual with a personal message expressing the chief executive's interest, concern, and expectations. The message may appear on the company letterhead or on an internal form if that is the normal method of executive communication. However, if the manual is designed to serve sales and public relations purposes, the communication should appear on the company letterhead.

A distinctive title page for policies, procedures, and instructions will help to distinguish the documents from less comprehensive material. The design should include the name of the company (or subdivision), the title or the topic being covered, a sequence number, the effective date, the person or the position of the person authorizing the publication, and the number of pages in the document. Use of the company logo often improves the appearance of the document. Blank title pages can be made in the form of duplicating masters, and centralized control over printing and distribution should be maintained.

To the greatest extent possible, manual documents should be complete within themselves. If many words with specialized meanings are used, a glossary of terms will be a useful addition.

3

Implementation

ONCE a program has been developed and documented, the implementation phase begins with three major kinds of activity: training personnel, administering the program, and amending the program to reflect new experiences and changing conditions.

TRAINING

The work of practically all production and maintenance employees—and most other employees as well—affects product quality. It is important that all employees support the principle that quality is an essential part of their job and the product.

Quality shortcomings can usually be traced to a lack of understanding or skills, or both. These deficiencies can be corrected, in most instances, by initiating suitable orientation and skill-development programs. When a new or enlarged quality assurance program is being launched, all employees should be given an orientation that will familiarize them with the importance of the program, its objectives, and its benefits. Once the program is established, all newly hired employees should be oriented, and occasional refresher courses should be given for all employees to sharpen their understanding and performance.

Employees, and especially production employees, should think of quality assurance as a normal part of their job, not as an additional burden imposed upon them by an outside organization.

Therefore, training in quality matters should not be wholly separated from other training related to job performance. From the worker's point of view, quality assurance should mean nothing more than doing the job right. Employees will need to know how to perform job tasks properly, what constitutes acceptable quality, and how it is determined. New employees should be informed about expected skill and quality levels at the time they are hired. This may require the company to develop job descriptions that incorporate quality expectations.

ADMINISTRATION

After the quality program has been developed, installed, tested and debugged, it must be maintained effectively. Engineering, design, purchasing, and manufacturing operations should be conducted within the framework of the quality requirements. Unobtrusively, quality considerations should permeate the organization. Quality regulating procedures should be followed and enforced, and work that does not meet specifications should be rejected and either repaired or rescheduled. The appropriate drawings, specifications, tools, and materials should be readily available and given out with the work assignments.

There are, in effect, three principal activities in a quality assurance program: establishing the proper conditions for good work; doing good work; and verifying that good work has been done. The quality organization is responsible for helping management establish the kind of working environment in which workers will want to realize established standards, and, in accordance with inspection and test requirements, it must ascertain that the completed work is up to standard.

Attitudes can significantly affect quality attainments, and a quality assurance program should not be allowed to drift into a game in which one side tries to get away with as much as it can while the other side tries to impose as many penalties as possible. The best way to preserve good morale and good quality is to make sure that (1) the required level of quality is realistic with respect to both customer needs and shop capability, and (2) effective inspection and test procedures are applied fairly, firmly, and uniformly.

KEEPING UP TO DATE

Probably nothing is as fatal to a good system as losing touch with reality. A system ceases to work when it loses credibility, and even the best systems gradually become obsolete unless they are periodically revised to reflect changing needs and conditions.

When a new system is being developed, there is widespread concern for its realism. And for some time after it is put into effect, there is an extensive effort to assure that the system works well in practice and that practice corresponds to the documentation. The subsequent deterioration of a system is a much more subtle matter, in which apparently harmless changes are made without considering the overall effect. At the time they are made, most of these changes will not seem important enough to warrant an overhauling of the system, but in time the particular procedures or instructions will be considered out of date and they will be ignored. If the attrition process continues long enough, it will destroy the reputation of the entire quality assurance program.

The dilemma of managing change in a quality assurance program is that such a program must maintain a fixed system while at the same time encouraging responsible change. Many a program has foundered on the assumption that its prime responsibility was to enforce a set of established rules. By maintaining such a rigid posture, the program insures its own ultimate demise.

The better course is to recognize at the outset that change is inevitable — and frequently desirable — and to plan for useful change as part of the overall quality assurance program. If this is done effectively, the problem essentially becomes one of perceiving when a formal change in the quality assurance program is needed, and then making the change in a way that enhances the value of the system. One difficulty in putting this into practice is that initially the need for change will often be apparent to the worker or foreman, whereas a request for a program change tends to originate at a higher level. As a result, time and credibility may be lost.

Two essential components of a grassroots updating plan are the ability of the quality assurance organization to respond quickly and affirmatively, and some means of recognizing the individual who makes the suggestion. There is no universal method for establishing these conditions, since they depend on the management style of the particular company. Part of the answer lies in setting up channels of communication that will bring the proposed change to the attention of the quality assurance organization as quickly as possible. In many organizations this can be accomplished through the formal suggestions system, although this will usually reward workers but not supervisors and managers, for whom some other form of recognition will be needed. In most cases management can reinforce participation in the quality assurance program by formally recognizing the employee's contribution and including, where appropriate, reference to that contribution in the employee's next merit review.

To summarize, effective implementation of a quality assurance program requires orientation and training programs to develop an awareness of quality assurance objectives and to develop necessary skills; a disciplined approach to maintaining quality standards; and systematic methods for introducing changes that will keep the program realistic and dynamic.

4

Maintaining the Program

AN established quality assurance program should be reexamined periodically to determine the degree to which it is meeting program objectives and current corporate needs. The examination should include an appraisal of the function with regard to its organizational structure, operating systems and operating effectiveness, as well as an evaluation of the degree to which the causes of shortcomings in quality have been identified and remedied.

AUDITS

An audit usually consists of a formalized inquiry, carried out and documented in a prescribed manner. An audit of the quality assurance function will particularly seek to measure effectiveness in meeting quality requirements that may have originated in company policy, contractual arrangements, or government regulations. The formalization of the audit is in part designed to enhance the credibility and value of the available information by regulating both the discovery of facts and the formulation of conclusions. For these reasons, it may serve as a basis for important management decisions and as a means of reassuring concerned individuals outside of the corporate organization, for example, major customers and regulatory agencies.

The design of an effective quality assurance program and its pro forma installation is no guarantee that the future quality assurance practices of the organization will be up to standard, or that the program will be revised and upgraded as necessary. An ordinary, statistical management report will not perform this task very effectively; an audit provides a more detailed and comprehensive look at the function.

To preserve objectivity, it is usually better to have the audit conducted by someone who is outside the quality assurance group, but who is familiar with it. The audit can be made by an internal auditor, by someone within the organization, or by a qualified outside organization. The focus of the audit may range from the general to the particular, depending on the corporate situation. A broad examination will be particularly valuable in the case of a program that has just been initiated, or in the opposite case, when an established program appears to be obsolete.

An effective way to begin a general audit of the function is to redefine or reexamine the quality requirements expressed or implied in customer orders, government regulations, and company policy. These requirements may well have changed since the program was first developed or even since the most recent audit was made. The quality assurance program must be broad enough to cover all quality requirements and narrow enough to exclude unnecessary ones.

The depth of inquiry of an audit, as well as its breadth, will vary with circumstances. For example, if a general examination of major functions reveals that good quality practices are routinely applied in the purchasing function, then a detailed examination of that function might not be necessary. However, it would be required if the overall good performance of the purchasing function was in fact concealing poorly performed subfunctions. Where standards of performance have been established and are valid for present conditions, most of the auditing can be accomplished by comparing the actual performance against those standards.

Although judgments by qualified people can play a significant role in an audit, they should not be a substitute for the facts; the development of an adequate data base is essential for auditing validity. To ensure that there will be enough data, the auditor (or auditing team) usually prepares a list of questions, and the answers to these questions will provide sufficient analytical details to make an objective evaluation of operating conditions and performance. These same details can help indicate where remedies are necessary, and provide a basis for delegating responsibilities so that the appropriate corrective action may be taken.

Where it is apparent that a program is generally effective but may have problems with specific procedures, processes, or products, an audit may be initiated in the particular problem area. Essentially, this type of audit involves a systematic examination of the problem area and a formal report of conditions and remedial opportunities. However, when focusing directly on a problem area, one should always keep in mind the possibility that the problem may be broader or deeper than it originally appeared. If necessary, the audit should be broadened to take account of all the causative factors and all the opportunities for improvement.

CORRECTING DEFICIENCIES

Theoretically, no quality failure should happen twice; it should have been corrected after the first occurrence. As a practical matter, mistakes do occur, and an error or failure rate based on past experience is sometimes used to establish a standard for production effectiveness. The value of such a test is directly related to its realism and rigor, but even if such rates have been set carefully, they often become unrealistic (and usually too loose) when conditions change. For example, if the expected failure rate was established when it was difficult to maintain required machining tolerances on the available equipment, and better equipment was installed later on, then that rate is no longer a valid measurement of quality performance. If failure rates cover too broad a range of activities, they will be of little diagnostic value because the good and bad conditions and performance will offset one another. Although failure rates are useful for some purposes, they have a tendency to limit quality expectations; in effect, they become expected quality losses, structured into operations.

A quality assurance program that does no more than filter faulty material out of the production stream is not a quality assurance program in the fullest sense. Sorting the work stream to keep a faulty product from reaching the customer is an important part of quality assurance, but it is only one aspect. Quality assurance implies *managed quality*, which includes finding ways to eliminate possibilities of error, and maintaining the proper operating conditions through such diverse activities as training operators, maintaining tools and gauges in good condition and supply, preventive maintenance, and good housekeeping.

Because responsibilities for quality performance are so widely distributed throughout the organization, difficult questions sometimes arise about who should undertake remedial work. For example, if quality performance falls off in a particular area, who should take the initiative in straightening things out? Should it be the person concerned with quality control, who obviously has specialized skills and concerns? Should it be the foreman, who is most familiar with the area and the people working there? Or should it be both? If the foreman and the quality assurance specialist cannot agree, who should prevail? This is another example of the difficulties inherent in line/staff relationships.

However, a generally acceptable way of dividing responsibilities is to give specific responsibilities to the individual who is in the best position to resolve the difficulties. For this purpose, a distinction should be made between random quality failures, which are presumably due either to human failure or to temporary conditions, and failures that form a pattern and are apparently caused by the failure of a quality assurance system. In the first instance, the problem is localized; it is essentially supervisory in nature. Unless the problem recurs with the same foreman, or in the same area, the responsibility of the quality assurance function is to notify the foreman and to provide help and counsel upon request. If performance in a particular area, or under a particular supervisor, is consistently below standard, the matter should be addressed at a higher level of management.

System failures, on the other hand, are a direct responsibility of the quality assurance organization. The matter should be thoroughly investigated in collaboration with line management, and steps should be taken to remove the cause of the defects. In some cases the proposed remedy will require the approval of other functions; for example, if new equipment is needed, a capital appropriation may be required. Approval of the appropriation may require action in several functions and at several management levels. Usually, the necessary coordinating activity should be undertaken by the quality assurance organization as part of its overall responsibility for the program. Once the needed changes have been identified, procedures for documenting, installing, debugging, and monitoring the improvement are substantially the same as those employed when the original program was introduced.

If the chief executive officer has made support of the quality assurance function clear and emphatic, there should be little difficulty in securing the cooperation of functional management. In the exceptional instance where line management is uncooperative, the effective operation of the quality assurance function may depend on the intervention of top management.

5

Contribution to
Corporate Results

VIEWING the quality assurance function exclusively as a cost center limits its potential usefulness. But if it is considered as a contributor to growth and profitability, its contribution should be ascertainable and, at least to some degree, measurable. Here the function is at a disadvantage compared with the rest of the organization. The costs of the quality assurance program can usually be readily ascertained, since they are made up primarily of expenses for personnel, facilities and equipment, and applied overhead. The benefits, while partially measurable by the cost accounting system, are to a larger extent realized indirectly; and when they are accounted for, they are usually credited to the operation or function in which they occur, not to the quality assurance function itself.

The most easily measured benefits of quality assurance are reductions in scrap and rework costs. In most instances, these savings will probably not equal the cost of the quality assurance operation. The larger benefits, which are more difficult to trace, relate to the positive influences on growth and profitability that the quality assurance program can exercise, for example, in increased sales that result from an improved quality image in the marketplace. Other potential benefits include reductions in field-service forces and lower reserves for servicing warranties and guarantees.

Where product recall can be mandated by the government or undertaken by the company, a reduction in historic recall rates can often be attributed primarily to quality assurance efforts.

Analyses of quality failures, prepared by the quality assurance function, can motivate quality improvements in the responsible operating areas, with results that register in the bottom-line figures. Examples include design modifications and improved purchase specifications that eliminate failure before it occurs.

An effective quality assurance program couples customer requirements with quality assurance practices and optimizes cost-quality relationships. Quality assurance can also play a significant role in the construction of facilities, the maintenance of equipment, and the support of a quality-conscious operating environment. The best quality assurance program satisfies customer requirements without waste or extravagance, is functionally efficient, and is effectively coordinated with the total operation.

A new quality operation, like any new operation, experiences initial costs that are higher than normal expenses would be. Although it may be financially desirable to charge these expenses against income in the expenditure year (to evaluate the cost-effectiveness of the quality assurance operation), it will often be better to apportion the initial costs over a period of years.

Viewing the quality assurance function solely from the point of view of cost may help keep the function lean, but in the long run profits will be reduced if the function is not allowed to make its full contribution. Admittedly, there is a substantial element of judgment in any evaluation of the function's "full contribution," especially its contribution to marketing results; but judgment always plays a major part in marketing decisions, and the contributions of the quality assurance system should not be excluded for this reason.

One way to approximate the value of the quality assurance function is to take the measurable savings resulting from scrap reduction and add it to estimates of the loss of profitability which might reasonably be expected if the quality assurance function were to be abolished. For example, if government regulations or contractual provisions call for specified inspection, test, or certification procedures, then abolishing the quality assurance function would deprive the company of its profits in that particular area. On the other hand, quality assurance costs can be considered opportunity costs. In addition, the effect of quality assurance activity on market performance can be estimated by marketing executives who understand the dynamics of quality in their particular markets. The relative cost of preventing quality losses in the factory and repairing defective products in the field can also be estimated. On the basis of such considerations, top management should be able to estimate the total contribution of the function to the corporation, and to decide to fund it at the optimum level.